COACHING CULINARY CHAMPIONS:

You, the Team & Competition

THIS IS A REVOLUTIONARY BOOK THAT WILL
CHANGE THE WAY CULINARIANS THINK
ABOUT COMPETITION!

Mareva Lynde & Frank Leake

COACHING CULINARY CHAMPIONS:

You, the Team & Competition

Mareva Lynde & Frank Leake

Seriouslyono Press seeks to Educate and Empower

Published by Seriouslyono LLC DBA Seriouslyono Press
PO Box 28097
Scottsdale, Arizona 85255

Limit of Liability Disclaimer
The authors and Publisher have used best effort in producing this work. They make no warranties with respect to the accuracy or completeness, and disclaim any implied warranties or fitness for a particular purpose. Material contained herein may not be suitable for your situation, and you should consult with a professional where appropriate. The authors and Publisher shall not be liable for any loss of profit, and damages to include special, incidental, consequential or other damages.

ISBN: 978-0-9799608-0-2
Printed in the United States

In memory of Joseph Amendola
The ultimate friend, mentor and teacher par excellence.
Your legacy lives on!

The Magic Thread.....

From your first day at culinary school, to the day you tryout for the culinary team, until the day you hang up your toque in retirement, you will be aware of the Magic Thread.... This thread is a continual search for excellence in all you do in the professional kitchen. It starts when you plan your menu, and select your ingredients, through the cooking and plating process, and doesn't end until your customer is satisfied.

This thread of excellence encompasses the essence of what it means to be a chef; the continual striving for the best you can offer. Anything short of that is a betrayal to you, the profession, and most of all to the customer.

Professionalism in the Professional Kitchen

To be vigilant about everything from buying the ingredients, to preparing and presenting the food for the customer, are the marks of a professional. Being a professional means taking care of the little things. Everything we do in the kitchen is about consistent quality, and making sure the customer is happy.

Being on time, and coming to work ready to work are the signs of a professional. It means not only doing your job, but also seeing what else needs to be done. Everyone in the kitchen works together, from the chef down to the dishwasher. The kitchen environment must be seamless, and efficient. You must come to work with respect and understand that it is a rigorous place and requires great discipline. When you understand that, and live that creed, only then can you call yourself a professional.

Chef Andre Soltner
Chef/Owner of the famed Lutece Restaurant, NYC, and
Dean of Classical Studies, The French Culinary Institute,
New York, N.Y.

You: The Team Member

You: The Team Member

At the end of this section you will be able to articulate your own strengths and weaknesses, and be able to create a personal plan to correct these weaknesses. As you and your class, guided by your coach, work through these pages you will see certain patterns emerging, and will learn a great deal about yourself, and how you fit into the culinary world. As you will see, a love of food, and a desire for stardom are not criteria for success in this very disciplined and structured environment.

There is ample space provided for note taking, and answering the questions posed. This book is a guide for your personal success as a team member and winning competitor. We want you to learn the skills of evaluating and questioning in order to better serve yourself, as you learn how to better serve your customer. Every day of your professional culinary life you will be a competitor.

Throughout the book you will find quotes from working chefs, both famous, and unknown to you, but revered by their loyal customers. What all these working chefs have in common is a solid understanding and ability to use, on a daily basis, the important precepts to culinary success. They all have acquired certain personal traits and management skills that propelled them to their respective successes. Information is knowledge, and knowledge is power.

1) What do you think is the most important personal trait you have that will bring you success in your culinary career?

_____
_____
_____
_____
_____

2) Are there any successes in your past that indicate you will be a successful culinarian?

_____
_____
_____
_____
_____

We want to provoke thought about the profession of food, and we hope to start a conversation between you, your fellow students/team mates, and coaches, will continue throughout your career as a culinarian.

The culinary artist that practicing professional, who takes raw ingredients and with a seemingly magical power creates "dishes" fit for the most astute tastes, was once a student like you. The path to culinary greatness is not he same for everyone, but no one can escape the rigors of the profession, and the training process. Whether you are a student, or intern, the road to greatness requires learning many things, some about food, much about cooking food, but most of all, learning about yourself.

Through conversations based on this book, while sharing your answers in class, you will examine important concepts that perhaps you didn't feel applied to you, and we hope you will learn to question your preconceived ideas about the process of food preparation in the professional kitchen.

The concepts examined in this book apply to all professional kitchens and chefs, no matter what country or language spoken in that kitchen. Universal truths about kitchen life will be explored. We promise you an exciting journey that just may prove to be transformative.

Humility.

To succeed in the culinary world you need to lose your ego, and be open to learning and discovery. Humility means you are ready for new experiences, and have left your preconceived ideas behind you. Without humility, and the ability to be receptive your road to "chefdom" will be thwarted.

Perseverance.

You will need to work hard at every stage of your career, and not give up. You will fail, and try again, but eventually you will master obstacles and succeed. There is no culinary success without daily perseverance. You must be humble and persevere to be able to call yourself a Chef.

Culinary School

Why are you in Culinary School?

_____
_____
_____
_____
_____
_____
_____

Why are you seeking a spot on the Culinary Team?

_____
_____
_____
_____
_____
_____
_____

Maybe you're thinking, "Aren't the answers to those questions obvious?" Actually they aren't. There are some things you should think about when answering these questions.

What is your relationship with food?

_____
_____
_____
_____
_____

Do you cook your daily meals?

_____
_____
_____
_____
_____

Was there a great cook in your life?

_____
_____
_____
_____
_____

Do you see this as a way to make a living?

_____
_____
_____
_____

What are your goals?

_____
_____
_____
_____
_____
_____

And perhaps most important of all—What do YOU bring to the culinary world? What makes you qualified to join the ranks of competing culinarians?

_____
_____
_____
_____
_____
_____

What do you think and feel about the concepts of professionalism, humility, and perseverance?

_____
_____
_____
_____
_____
_____

Being a Chef

What does it mean to be a chef?

_____
_____
_____

The title and privilege of being a Chef is something that you must earn. Getting a certificate or diploma doesn't make you a chef. Being a chef is a lifestyle. Hard work, sacrifice, and understanding that there are no shortcuts, will eventually lead you to the rank of a Chef.

Have you thought about what you need to learn in order to become a Chef?

_____
_____
_____
_____
_____
_____

How will you handle failures?

_____
_____
_____
_____
_____

How will you handle success?

_____
_____
_____
_____
_____

There are No Shortcuts

If you are looking for shortcuts and clever ways to cut corners you don't belong in culinary school, and you certainly have no business attempting to enter competition. Sadly there are always competitors who try to cut corners, shirk work, and skirt around the rules. Tricks to save time don't work, and is a form of cheating. In competition you would be cheating yourself, and letting your team down, if you attempt to cut corners. If you spend your energies in trying to circumvent the rules, the standards, and then hope to produce a quality product you will be disappointed. It is better to spend your energy on winning by the rules. If you pay attention to technique, and teamwork, you just may find yourself on a winning team. Remember, this is a team event, if you cheat you bring down the whole team. Do your job as well as you can, and leave shortcuts to the losing team.

1) Do you think there are any shortcuts that are okay in competition?

_____
_____
_____
_____
_____
_____

2) You really want to win, so what is so wrong in producing badly diced veg, when it's faster, and no one is likely to notice?

_____
_____
_____
_____
_____
_____

3) Your skills are good enough, but your fine dice isn't perfect. It's so small who will notice when it's cooked? Aren't the judges just hung up on small details?

_____
_____
_____
_____
_____

What About Fame & Fortune?

What if we tell you that you will spend most of your life knee-deep in potato peelings and onion skins, in a perhaps not idyllic environment, surrounded by colleagues and co-workers who do not have your vision, and that you will be required to turn out meal after meal, to customers who may or may not appreciate all your effort and talent? Sound like heaven or hell; neither it's just the reality of the professional kitchen. Are you nodding your head in agreement, or shaking your head in disbelief? Or do you see it differently?

1) What do you honestly think about the last statement? Don't hold back, the only right answer is the one you will need to live by, and believe. This journey is yours, and you will have to live with the result of your choices.

_____
_____
_____
_____
_____
_____

2) Do you think you can create a more harmonious environment where you work? Why, and how will you do this?

_____
_____
_____
_____
_____
_____

3) Where did your view of the professional kitchen originate? Were you a customer, did you see a cooking show, do you work in a kitchen, or do you know someone who actually works in a kitchen?

_____
_____
_____
_____
_____
_____

4) What about your favorite cooking show stars, the newest foodie on the block, the one with his own cookware, or the one with his own sauce line? Is it all about the sizzle, and if you look closely there is no steak?

_____
_____
_____
_____

5) Why are you here today, with this book in your hands? If you need to think too long, or are unsure, maybe you are not in the right place.

_____
_____
_____
_____
_____

Honesty

Honesty should be in everything you do, and cook, and serve to your customers. Honesty is about reality. You need to know yourself, and your limitations. Your plated food speaks to who you are. Quality ingredients, cooked with care, and respect for the food, yourself, and your customer is paramount.

Respect

You must respect yourself first. Respect must emanate from you to your colleagues, and to the ingredients you will use in preparing meals. If you don't respect the food, you are telling your customers that you do not respect them.

Self-Evaluation

Every day you must honestly evaluate that day's performance. It doesn't matter if the Coach didn't see, or your teammates don't care. You are in the race by yourself, and you will win or lose your potential talent and future greatness if you do not use honesty and respect as tools to measure your achievements as well as failures. Your future depends on your ability to care enough about yourself to do so everyday. Remember to critique and do a daily analysis of your work.

Honesty/Respect Self-Evaluation

You eat food everyday, probably three meals a day, and some snacks. How much thought is attached to that activity, and to what you are actually eating?

_____
_____
_____
_____
_____
_____

Have you ever thought about your relationship to food? Do you cook your own meals? Are you eating to savor, or fill your belly?

_____
_____
_____
_____
_____
_____

Do you shop for the ingredients? What are the criteria you are using for selecting the ingredients? Price, size, color, freshness?

_____
_____
_____
_____
_____
_____

Do you eat cold cereal over the sink, grab fast food, or do you walk past an eggplant and immediately think of five ways to cook it this evening?

_____
_____
_____
_____

These questions strike at the heart of what it means to be part of the culinary world. Honesty, respect and self-evaluation will give you a better understanding of your ability to better pattern your career in a successful way.

Assessment

When you do something no matter how small, or insignificant like getting into your uniform, you are creating an environment that will be assessed by others. You will be judged, appraised and actually you will be defined by that act. A sloppy knot in your neckerchief, a stained jacket, and dirty fingernails will all declare to the world that you have lowered the bar on what is acceptable.

In everything you do, you are expected to stand back and evaluate and assess what you have done. We expect you to be honest, and evaluate against established criteria all that you do as a culinarian.

When you set up your mise en place in a disorderly fashion, when you serve chicken that is underdone, or overdone, or when the plating is messy, you are shouting from the roof tops that you do not care, and you have neither respect for the food nor for the customer. It also says you don't care about yourself.

1) How are things assessed? What do you think is more important, appearance or good intentions?

_____
_____
_____
_____
_____
_____

2) Do you think a chef will have a positive opinion about your sloppy plating, even if your coach/ instructor was critical?

_____
_____
_____
_____
_____
_____

3) Why does your coach/instructor make you do things over and over again? Do you feel that he/she is just sadistic or has different taste?

_____
_____
_____
_____
_____

4) Everything you cook is so delicious, but your grades don't represent that, and your coach keeps telling you to do it again? Does it really matter if your vegetable cuts aren't uniform?

_____
_____
_____
_____
_____

Personal Assesment Review

1) What are your broad culinary goals?

_____
_____
_____
_____
_____
_____
_____

2) Have you clearly defined your objective: French restaurant, caterer, or haven't you thought that far ahead?

_____
_____
_____
_____
_____
_____

You need to take some time now to think about why you are in culinary school, and where you see yourself next year, in five years, and in ten years.

3) Why should you be selected for the team?

_____
_____
_____

4-If you have a specific goal, what do you need to do to achieve those goals?

_____
_____
_____

5-What are some of your shortcomings that could stop you from getting to your goal?

_____
_____
_____

6-What skills do you have that will help you achieve your goal?

_____
_____
_____

7- What should you do if you have difficulty answering the above questions?

_____
_____
_____

8- What resources are available to you that could help you answer these questions?

_____
_____
_____

Being There

Being there means, being on time, in a clean uniform, ready to work. That means you reviewed the day's work, and required outcomes, and output before you got to class, or practice. On time means getting there 5 minutes early. Your life is overflowing with commitments, like family, work, and socializing, and you feel torn in many directions, so you are skimping on the culinary side. Make no mistake, being there for culinarians is the first step to any measure of success.

Some of you may have difficulty waking up in the morning; buy two alarm clocks. Put your uniform out the night before. Make sure your tools are ready to go. My dog ate the homework doesn't work anymore. Make a checklist, and use it!

You're already late if you are arriving five minutes before your class, shift or practice begins. Being there means you are ready to reach for success, and you are ready to listen and take direction. You are the master of your time, and only you can determine your level of success. Just showing up and hoping to figure out what needs to be done on the fly doesn't work, and you probably know that by now. You are smart, and quick, and a fast learner but the learning starts before you arrived in the morning.

Everyone is different and has different outside responsibilities, but all of you share one thing, the desire to succeed in the culinary world. Not everyone will be a TV star, that is certain, but all success comes from your ability to figure out how to do the work necessary, so that early every morning you are a "learning machine." Your coach is there to guide you and your team members, your chef/instructor is there to introduce and teach, your job is to be ready to receive and learn.

1) Do you think learning on the fly is acceptable?

_____
_____
_____
_____
_____

2) What do you think is your responsibility vis a vis your instructor, your team mates, your classmates?

_____
_____
_____
_____
_____

3) What if you have managed to slide by so far, not very well prepared but still holding your own? Is that acceptable? Is that honest? Do you think you can build a culinary career in this fashion?

_____
_____
_____
_____
_____

The Art of Practice
in the Professional Kitchen

You are what you practice. What you practice is what becomes your habits. It's not really practice that makes perfect, it's perfect practice that makes perfect.

The reason that people get good at what they do, is that they have a relentless pursuit in doing something over and over again and doing it a certain way with vigilance.

You won't get good at something if you do it a few times, or you read about it and understand the concept, you have to get in there and repeat it over and over and over until your body responds without effort.

People think that they just need to get past being the asparagus peeler or the potato peeler so they can move on and move up, but moving up is not about being "chef." It's about what you become on your journey to being "chef." It's really about peeling the asparagus and peeling the potatoes.

I you haven't done it at least 100 times, PROPERLY, then you haven't practiced it enough.

Chef Keoni Chang
Corporate Chef
Foodland Super Markets, Ltd.
Honolulu, Hawaii

Reliability in the Professional Kitchen

Reliability is an essential to success in life. Punctuality, showing up on time, putting out consistent, quality product each time you serve a dish, commitment to excellence in all that you do, all essential and directly connected to your reliability.

As a cook and a chef, it is imperative to have timelines engrained as part of who you are and as part of your normal approach and routine to your life as well as your career. In other words, practice and commit. The "three strikes you're out" or in this case, "three late's and you're gone" are communicated right up front. There is no excuse for not being on time. Excuses are just that, excuses.

Communication skills are essential and define how reliable you actually are. Not returning phone calls and emails are real pet peeves for me. No response tells me allot about a person. This lack of communication instantly sends a red flag, "non-reliable."If a person won't communicate in a timely manner, they are not going to be reliable when crunch time comes to putting out that meal or that party on time.

For me, reliability in the workplace begins before the technical work actually begins. When that cook walks in the kitchen they better be ready to go. That means full uniform (cleaned, starched and pressed jacket, trousers, apron), clean toque, clean socks, proper shoes, hair (clean and in place), bathed, etc. If not, reliability is in question and if that be the case, trust is immediately broken. Members of a classic brigade would never insult the chef by mis-representing their reliability by wearing a dirty uniform.

As a young cook in training I always showed up on an average of 3-4 hours early every day. This provided me time to catch up and to learn other things from the staff that I would not normally have time to learn. I became more capable as part of this team.

I would work the graveyard shift, go home at 7AM and return at 2PM finding that the guy who was scheduled to come in at 7AM when I left either never shows up or comes in late. One way or the other, I'm done if I haven't walked the extra mile in prep. It's a chain reaction of events. One late call or no show affects the entire service and every guest. This is unreliability. Partnership and the concept of team is now gone.

Reliability means that you can count on someone to be there to perform and to complete everything at the highest levels. Quality and quantity levels must be accomplished for survival of the operation and of the team.

At 20 years old I was the youngest apprentice to take over the line station of saucier in the hotel where I was training and working. I worked as a roundsman to learn all the stations. That meant that the team of cooks and chefs I was learning from cam to depend on my reliability to be there to perform and to learn. Reliability is all about ethics, morals and values. These I got from my family, they were engrained as I grew up. By the way, I'm it, there are no former chefs or professional cooks in my family. I went out there and chose the tough life, my decision. Therefore I had to buy into everything that defined what becoming a cook and chef meant. I had to first be reliable to myself, then to others.

I find that I can see reliability as I interview and hire young cooks. I look for less or no skills so that I have fewer problems in breaking bad habits. I listen for philosophy and attempt to learn more about the person and have them show me the technique. And I learn allot about how a person shows up for the interview, even a dishwasher needs to be appropriately dressed and prepared to interview.

When I expedite, I methodically return plates to the kitchen, even if the plate is near perfection. This keeps the staff on their toes. They inspect, they must identify what needs to change, they become motivated because they must buy into the perfection required by the guest and of the team.

Someone doesn't live up to all these qualities, get rid of the dead weight. Dead weight = unreliability = poor partnership.

If I had to do it all over again I would not change one thing. The lessons learned where hard but profound as I continue learning and growing as a chef.

The concept of reliability evolves with years in the business. You build both internal and external teams to back you up. Mentors and friends are a phone call or an email away when you're in trouble, or as we so often say in our business, "in the weeds." I can pick up the phone and call Paul Prudhome and ask for help. You develop a core of people who you can depend upon and in turn they depend on you. I am always there when a call comes in. There's no expectation on the part of the asker or the provider. You don't help someone out because you want something in return, you help because you care. You've lived your life caring and living your reliability and dependability.

I keep learning, I keep caring, I keep sharing, I keep trusting, I'm reliable and that won't change.

Chef Jeffrey Mora
Corporate Chef and Owner, Metropolitan Culinary Services Inc.
El Segundo, California

Do Your Job

Attending classes, or preparing for competition is part of doing your job. Just showing up, with a bright smile won't cut it. Preparation for the day is part of your job. If you've made it to the team, your job is to harness all your physical, psychological and mental powers to get the job done. You can't hope or expect the others to sweep you along. All of you need to show up strong, ready to handle all adversities, and mishaps. Getting the job done requires many skills, emotional maturity, and the ability to work through all the things the day will bring. Everyone needs to be as prepared, polished and accomplished as possible in order to achieve success.

1) One of your teammates is good at logistics, so you've come to depend on his/her talents to make sure all ingredients are ready. What if that teammate is sick, or goes out of town. Who will take over?

_____
_____
_____
_____

2) Your friend enjoys examining the produce in the morning, and he's been gathering the fresh ingredients for both your mise en place. You've allotted yourself the dry ingredients, and herbs. Do you think this is a fair exchange?

_____
_____
_____
_____

3) Do you consider this kind of behavior honest?

_____
_____
_____
_____

4) If you have no experience making sure the right ingredients are available when you need them, and you have little to no experience selecting your raw ingredients, what are your plans to learn how to do this?

_____
_____
_____
_____

5) What do you think is the most important thing you need to know before you start each day's work?

_____
_____
_____
_____

Yes, Chef!!!

You have probably heard that famous phrase, " Yes, Chef," and maybe thought, "What's that all about?" Maybe you giggled, and wondered who would say that. Since you've been in class, and are on the team or trying out for the team, you have probably said, "Yes, Chef," a few times yourself.

What does that really mean? Is it a term of endearment, or a polite way to acknowledge your instructor? Neither, it means you are acknowledging that what your instructor has said is the standard, the only standard, and that there is no discussion. It's not a democracy in the kitchen, you do not get a vote, and that there is no personal opinion that counts except for the chef's opinion.

That may seem a bit harsh but the professional kitchen is a disciplined environment that must produce meals of quality on a consistent basis, not on a whim, and you must produce quality even when you are broken-hearted, tired, sick, or feel that you need a vacation.

1) **What do you feel about the phrase, "Yes, Chef?" Do you think times have changed since Escoffier, and it no longer holds true?**

_____
_____
_____
_____

2) **Do you think your opinion should hold some weight?**

_____
_____
_____
_____

3) **Do you feel that your creative instincts are being smothered by all the 'Yes. Chefs" acknowledgements?**

_____
_____
_____
_____

In the professional kitchen, "Yes, Chef" is the rule. Congeniality, and camaraderie notwithstanding there is really only one chef. Mutual respect doesn't clash with "Yes, Chef", and if your opinion is asked you may give it, but democracy isn't what the professional kitchen is about. In order to succeed you must remember this culinary truth.

4) **What do you feel about this statement? Does being a small cog bother you? Do you think there is a better way to run a professional kitchen?**

_____
_____
_____
_____

Know Your Place

Everyone comes to the culinary world from a different place. Some grew up in a strict home; some had parents that were admirers of a looser form of discipline. Some people went to public school, others had their hears boxed by the nuns. Some people had careers; positions of authority, and some people came to culinary school, straight out of high school. It really doesn't matter where you came from, everyone starts at ground zero. You will be a sponge, you will soak up everything you can, you will say "Yes, Chef," even when you aren't sure why you are saying that, and above all YOU WILL KNOW YOUR PLACE!!!

That may seem a little harsh, because you feel you have talent, and a great future a head of you. Sometimes you think your chef/instructor is just plain wrong, or maybe stuck in the 1960's. This is 2008 for goodness sakes. You need to understand that though time moves on, and we put foam on foods, and sushi is found in every supermarket, the fundamentals haven't changed since Escoffier. No, you're not in France, and you have no intention of cooking French food, but everything flows from the Gold Standard. Maybe you have your sights on a small sandwich shop that you will run with your friend, or maybe you want to be a corporate chef, it still stands, that there are not two ways to do it. There is only your chef's way. There is a reason that chefs see things a universal way. Whether you travel to England or France, the rules are the same, and your chef/instructor is attempting to show you the Standard way of doing, seeing and appreciating things. Your job is to learn as much as possible from each chef/instructor, all the while remembering to be humble, and to evaluate your outcomes.

1) Do you think you can by-pass many things your chef asks you to do? Describe how you would do things differently.

_____
_____
_____
_____
_____

2) If you are an older, or second career student; do you find it hard to accept the Chef as the authority figure?

_____
_____
_____
_____

3) You grew up in an environment, where your parents valued your opinion, and you feel your chef/instructor is too dogmatic and doesn't respect your opinion. What can you do about this? Should you do anything about this?

_____
_____
_____
_____

4) Where is your mind focused while in class? How about in practice for competition?

_____
_____
_____
_____

Open to Ideas & Learning

When you have finished your formal studies, the real work of learning will begin. Every job, or position you will hold is filled with opportunities for you to learn. If you feel that you have learned enough, you will not flourish and your career will come to a standstill. Keeping your eyes open, and being open to what others say, and do will keep you on a path that can only lead to success. Being able to see food through the eyes of someone else will permit you to understand other cultures, and learn about new possibilities. Food is as old as mankind, yet there is always something new to create.

1) Do you have any restaurant experience that has opened your eyes about a certain ingredient?

_____
_____
_____

2) Do you have a favorite cookbook that you use constantly? What is it, and what kind of food are you cooking?

_____
_____
_____

3) Is there an ethnic food that you eat, but never cook? The next time you eat in this restaurant, deconstruct the food, figure out the spices, consider what cooking method was used.

_____
_____
_____

In all your activities that are food involved, be there, be involved, and learn something new.

4) After class today get together with 3 other classmates, and pick two ingredients, such as tomatoes and onions, or corn and butter, and talk about what different cultures do with those ingredients, and then come up with a recipe for those ingredients. You will see that the ideas will flow, and you may even learn a new way to use tomatoes.

_____
_____
_____
_____
_____
_____

Artistry in the Professional Kitchen

Culinary schools in the U.S are fast becoming very popular like the fast food chain McDonalds all over the world. A lot of students think that after they graduate, they become instant "CHEFS". Unfortunately, they need to understand that it's a learning process. The students need to ask themselves, what's their experience? Do they understand the industry? Do they know what's going on..........?

Some people can't handle more than 8 hours, they can't work O/T, they don't want to work weekends or holidays, or that it's not enough money for them to do the job, but yet on the flip side they acquire satisfaction for themselves as well as their family members and of course manage to receive a Degree.

The Culinary industry is a lot alike the farming industry, a lot of hard work requiring a good attitude to be able to tackle long periods of time and stress in order to grow mentally and professionally. The final outcome will eventually reap its benefits and results. It is at this time when one asks himself "Am I satisfied with what I've accomplished in the past".

While in Las Vegas, I was honored to be able to join the competitive culinary team, Las Vegas Team U.S.A. Along with my team, we never turned down a challenge, especially if it provided more opportunities to bring my culinary knowledge and skill to another level. The extensive and diverse training I have received both in study and on-the-job, allows me to prepare dishes of virtually any realm—classic European to eclectic fusion. One may think that if someone surrounds themselves too much in one thing, it may make that person one-dimensional and lose touch with the world around them. Having complete focus in my craft is of the utmost importance. Just like the teachings in Tae Kwon Do, one must respect the power and energy he brings to the practice. Having earned black belts in Tae Kwon Do and Hapkido, I continue to study and practice these forms of martial arts, not to fight, but to better ground myself and better evoke patience and creative, positive energy. I strongly believe in a saying by Bruce Lee, "Be limited to be unlimited," and my own saying, "The impossible can be possible."

Our lives are like the body builder in training. If you work hard and do it right, you will see the results. But if one tries to cheat like the steroid user, one will eventually get hurt and hurt the body system. We have to talk to ourselves, what is our goal, what do you want in 5 years? Are you meeting the goals yet?

Chef Richard Han
Chef Garde Manger
Pechanga Resort and Casino
Temecula, CA

Creativity & Artistry

Have you ever eaten a Cobb salad, or surf and turf? Do you remember when Martinis were derigeur, and then fell out of fashion? There is fashion in food as well as clothing. Pizza has gone upscale, and martinis are in again, although this time in flavors like "green apple."

1) **What have you eaten lately that was beautiful, or artistically plated? Did it work? Why? If it didn't succeed, where did it go wrong? What did you learn from this experience?**

_____
_____
_____
_____
_____
_____
_____

What's hot today, maybe be old tomorrow, but your food must adhere to certain principles. Poor quality ingredients, badly cooked will never be good, even if you place dots of balsamic vinegar around the plate or swish corn foam on the side.

2) **What do you think is the underpinning for creative food? What do you need to learn in order to produce creative food?**

_____
_____
_____
_____
_____
_____
_____

Creative food is a respect for the raw ingredients, and the ability to visualize, smell and taste what the possibilities are. Soup doesn't have to be boring. You could cook soup everyday of your life and never repeat the same one. Creativity is about understanding the limitless possibilities.

3) **Do you think every chef has the potential to see all the possibilities?**

_____
_____
_____
_____
_____

4) **What can you do to make yourself a more creative chef?**

_____
_____
_____
_____
_____

Artistry can be about plating and presentation, or about finer food pairings. Fish soup is available in almost every culture, yet Bouillabaisse tastes and looks so much different from Vietnamese Catfish soup. Artistry is about clarifying cultural strengths and creating new combinations. Artistry comes from hard work, and a deep knowledge of a culture's food. Combining, and mixing, and borrowing from different cultures is much harder than it seems, when it's on your plate.

5) What would be the biggest impediment for you in having to cook a Peruvian dish if you aren't from Peruvian ancestry?

_____
_____
_____
_____
_____
_____

The Customer

"This would be a great job except for those darned guests!"

Who are these guests anyway?

They work longer hours than they would like to, they make more money in a year than their parents made in ten, they value their time with their families, they are willing to spend a substantial amount of money so long as there is perceived value, they are more widely traveled than any previous generation and they know more about good food and wine than most of our average employees.

In the last 20 years there has been an explosion in the number of Luxury resort hotels available for the affluent traveler to choose from, so the competition for their business is stiffer than ever before. The only resorts that will be successful in the long term are those that can satisfy all of the needs of their guests. It is essential in this arena to satisfy not only the verbalized guest preferences, but to anticipate their needs and to fulfill them before being asked. "Make it happen before they ask" is the road to success. When you really get down to it, isn't that what we really mean by hospitality?

If you invite some of your best friends over for dinner, I would imagine that you try to figure out what they like, if they do not eat something, if they like red or white wine. Do you meet them at the front door with a smile? do you offer them a cocktail when they arrive? and when they are leaving do you walk them to the door and stand there waving as they drive off? If not then I expect you probably do not have a lot of friends. "It starts with the customer," is one of the core values of The Greenbrier and its parent company, CSX. In order for any company to translate that philosophy into action they must have access to a staff that lives and breathes this philosophy. They need to hire the attitude not the training, that is to say, that we must have access to people who are willing to be trained. Training can only be transposed onto a willing participant with the right attitude, passion and work ethic. The development of these qualities is something that has to start early, and for the most part happens in schools and colleges all across the country. It is the teachers, culinary instructors and professors who are charged with this task, and who for the most part keep this industry supplied with the right talent. I would like to take this opportunity to thank all of the teachers and mentors for their unwavering commitment to training the attitudes and passions of our hospitality professionals of the future.

Chef Peter Timmins, CMC
Executive Chef, The Greenbrier
White Sulphur Springs, West Virginia

Core Values

Are you able to articulate what your core values are? It is very important for you to understand that who you are is what you produce. In order to give your customers the freshest, and the tastiest of meals, you need to always go back to your core values, which should include: honesty, integrity, ethics, striving for excellence, respect for the food, and the customer.

Your tomatoes aren't looking so good, and the fish has an off smell, but you decide to make a Mediterranean dish using both for tonight's special. You know that if you season it with a heavy hand most of your customers won't realize, and the few that do can be offered another dish.

1) Are you comfortable with this kind of decision?

_____
_____
_____
_____
_____
_____
_____

2) What could have been done so that you wouldn't be in this position?

_____
_____
_____
_____
_____
_____

3) How much does the bottom-line get in the way of acting with integrity?

_____
_____
_____
_____
_____

4) Some of you may be asking what is wrong with serving this dish, since it's possible that no one will get sick from it? What is your responsibility toward your customer?

_____
_____
_____
_____
_____

Staying Healthy & Stamina

Staying healthy, in both mind and body is so important, that if you neglect this area your career can be derailed, or come to a grinding stop. What other career choice requires brawn, the ability to lift heavy bags of root vegetables, and at the same time be expected to have the artistry and ability to delicately plate the latest in food trends?

Maybe you're thinking that a few late nights with your buddies is good for your morale, and that we just don't understand. We do understand, because late nights, and early morning job commitments don't work. At first you will probably get away with it, but if you set that up as a pattern, you will see your health suffer, your mind will wander, and your creativity will grind to a halt. Kitchen life is not for the weak, and not for the uninformed.

The life of real chefs, working chefs, is predicated on good health, clear thinking, and having great stamina. Chefs work in many arenas, from corporate to great restaurants with many patrons every night to the chef/owner who has 16 seats, and cooks their style of home-style food. What all these chefs share everyday is long hours, which is difficult, but added to that the quality of everything that leaves that kitchen, from the first dish to the last at 11pm must reflect that's chef's version of perfect. Every dish that leaves the kitchen reflects who you are, and if you are too tired to care, or too exhausted to meet the criteria you won't be a success for very long.

1) Clarity in thinking, and responsible thinking will help to further your career. Sloppy thinking, when you're exhausted from partying will bring your career down, if not destroy it. What do you need to do to stay sharp and focused?

_____
_____
_____
_____

2) How much sleep do you think you need?

_____
_____
_____
_____

3) When should get-togethers and partying happen?

_____
_____
_____
_____

4) Do you get enough exercise?

_____
_____
_____
_____

5) How are your eating habits?

_____
_____
_____
_____

6) Are you organized enough to get to bed on time, and know your knife kit and uniform are ready to go first thing tomorrow morning? Or do you toss and turn knowing your jacket is filthy, and even your apron won't hide the stains. Your station is towards the back of the room, and maybe Chef won't see. Yet you know Chef sees everything, and you'll probably be embarrassed in front of the entire class, and then you'll have to pretend to be so very cool, and laugh it off.

_____
_____
_____
_____

Stamina is the ability to make it through no matter what. Stamina comes from being healthy, staying focused, and being prepared. When you have stamina you can get through the day even if you have a cold, your back hurts, and you're mad at your Chef, or fellow culinarians. Stamina is the deal breaker. Without it you won't go far. But, first you need to be clear-headed, healthy, and prepared.

1) Do you notice that towards the end of the day, your mind wanders, and your back is starting to hurt? Are you popping over -the -counter pain relievers, and drinking too much coffee to make it through the class, practice or shift?

_____
_____
_____

2) Stamina is a mind/body phenomenon, so you need to be as healthy, and clear-headed as you can be. What things are you doing currently to build your stamina?

_____
_____
_____

3) You're still relatively young, and think this doesn't apply to you, but after a night of partying, many hours of missed sleep, and too much work, do you think you can still make important business decisions?

_____
_____
_____

4) Have you ever talked with working chefs about their life-style? If you haven't this would be a good time to make friends with a working chef, and ask them about how they continue to have stamina, and thrive in the hectic business that is the food business.

_____
_____
_____

Understanding the Standard

Small dice is small dice in any kitchen where the chef is classically trained. Whether in Hungary or Spain, California, or New York, small dice means only one thing. In the professional kitchen there is a universal language that is understood the world over. Escoffier is long gone, and new heroes in the culinary world are coming over the horizon. Yet everything Escoffier stood for still stands today.

Everything stems from Escoffier and the classics. Everything new is a riff on the old. Interpretation is homage to the past masters, yet nothing is really new. Integrity and consistency still matter. No one can count themselves a success if they succeed one day a week. You have to cook, and plate and serve at the same level everyday. Anything else is below the Standard.

The Standard is not negotiable. Caramelized onions, means only one thing, and it is a certain color, and consistency. You do not create the Standard, but you will interpret the Standard, and hold it as a mirror to everything you do in the kitchen. The Standard covers everything from cleanliness, to cooking. The Standard is about rules and principles, and separates the chefs, from the hacks. Understanding the possibilities of an onion, permits us to cook Soupe a L'Ongion, and fried onion rings. The Standard defines what it means to be sautéed, or fried, or sous vide.

Maybe you're thinking that you want to open a Mexican Restaurant, or a Sushi Bar, and you think the Standard doesn't apply. Certainly Escoffier didn't know that sushi was to become so popular. Yet, the Standard still applies. Consistency cannot be accomplished without understanding the Standard.

1) Do you think using a recipe faithfully, yet not following the standard will achieve a good dish? Why do you think so?

_____
_____
_____
_____
_____

2) If you are a chef/owner would you make sure your team understands that there is no deviation from the standard?

_____
_____
_____
_____
_____

3) Do you understand that adhering to the Standard will have an impact on your career? Why? If you don't agree, why not?

_____
_____
_____
_____
_____

Expectations: Understanding What is Expected of You

"Expectations" is a concept that covers information that goes to and from you to your Chef/Instructor and classmates.

That are many things you expect to learn while in school, and hopefully by competition time have you will have absorbed/assimilated many of these things. You expect to learn about or improve your knife skills, learn the intricacies of sauce making, cooking in large-scale mode, and graceful plating.

1) What are some concepts that you have learned so far that you were unaware of before?

_____
_____
_____
_____
_____
_____

2) Have you learned anything from your classmates that you were unaware of before?

_____
_____
_____
_____
_____
_____

3) Are there things you had hoped to learn that you haven't? And if so, what are they?

_____
_____
_____
_____
_____

4) If you feel that there are concepts both physical, like techniques, or philosophical, like concepts, that you are unsure of why haven't you sought information from your Chef/Instructor?

_____
_____
_____
_____
_____

Understanding What the Chef/Instructor Expects of You

Your Chef/instructor expects you to listen, to learn, to become the best you can be in the professional kitchen. They expect you to ask questions in order to learn better. They expect you to seek help when you haven't mastered a technique. You are also welcomed if you are seeking knowledge. What is not welcomed is questioning the Standard, or the Chef's instruction.

1) Do you think your Chef/Instructor is a stickler, and should let some things slide? What things are these, and why?

_____
_____
_____
_____
_____
_____

2) Do you sometimes feel you are being criticized unfairly, or are you feeling insecure when being critiqued? What do you think the Chef/Instructor is using as the yardstick to evaluate your efforts?

_____
_____
_____
_____
_____
_____
_____

3) In the workplace kitchen, what do you think will be expected of you?

_____
_____
_____
_____
_____
_____

4) Why should any workplace accept below standard performance?

_____
_____
_____
_____
_____
_____

Ego: Feeling Like a Star

Do you find yourself admiring your chef's jacket more than once a day? Are you pleased with the knot in your neckerchief? Do you and your friends discuss the merits of fine Egyptian cotton neckerchiefs and jackets, yet you are still in Intro to Fundamentals? Do you feel your knife kit doesn't suit your personality? Are you too concerned about your image?

It is important for your uniform to be clean and wrinkle free, and your nails short and clean. Your hair should be styled for practicality, and to enhance your professional image. Your head covering, should do just that, cover your head. A jaunty cap sitting on the back of your head isn't doing the job, that of keeping your hands out of your hair, and your hair out of the food. The job at hand is to learn as much as possible about the selection, preparation, and plating of food. Your customer's happiness is your main objective, and the only star in the bunch should be your food. If you are hoping for a show, or a career where your uniform sparkles, and your long red nails toss the salad, you may have chosen the wrong profession.

Understanding what everyone's role is will better help you accommodate to the reality of the professional kitchen. Your Chef/Instructor is there to guide and teach you, and your job is to learn as much as possible. It's good to be happy with each day's accomplishment, but it is equally important to understand that you are moving toward your goal, and that even after graduation you will still be learning. When a Chef stops learning his career comes to a standstill.

A healthy ego means you understand what you have accomplished, but it doesn't mean you have arrived. The life of a Chef is about the journey, discovery, and respect for the food.

1) Everyone says you should own a restaurant, so why even listen to anything the instructor says?

_____
_____
_____
_____

2) You're planning on opening a vegetarian restaurant, so why learn anything about chicken?

_____
_____
_____
_____

3) You feel that you are really talented, and just need that diploma to launch yourself, and your instructors are holding you back by teaching classic methods. What should you do?

_____
_____
_____
_____

4) Good knives are essential, but does a beautiful knife carrier make you a better chef?

_____
_____
_____
_____

Ego

My recipes? Of what value is it for me to teach you my recipes, as it is more important and valuable to you for me to teach you the methods of recipe. Many say that "My" recipe is better, but it is how we put it together that matters. This again takes us back to the method.

We have accepted the teaching profession to educate young culinarians with fundamental skills that will be the building block of their entire culinary career, and of value to the many chefs they will work for.

The true measurement of your success is when a young culinarian takes their first position and the chef comments that they have great fundamentals. That is success.

Why do we need to teach them "My" recipes?

Chef Thomas B.H. Wong, CHE ,CEC
Chef Instructor
The Culinary Institute of America at Greystone
St. Helena, California

Insubordination

You're not in the army, but the life of a culinarian is somewhat militaristic in the sense that there is structure, and hierarchy in the ranks. There is no democracy in this scenario. The Chef is the boss. You have two choices, follow, or get out of the way. Your opinion is better kept to yourself, unless asked. The Chef is in charge, and expects compliance. The kitchen cannot do what needs to be done if there is discussion and questioning among the ranks. You can certainly ask questions that will help you to better accomplish, or complete tasks, but you cannot question the task.

Every Chef has a vision of what his cuisine is. Technique is expected from you, and the vision belongs to the Chef. He/she expects small dice, and caramelization to be on- point, not approximations. You are expected to be there to work, accurately, cleanly, and without comment. Unless you are asked for your opinion, you don't have one. This may seem a little harsh, but by this time in your culinary career you understand how much work goes into producing meals in a commercial establishment. There is no time for personal moments of expression or tantrums. The chef has the ultimate responsibility for every plate that leaves the kitchen. The brigade concept still holds, and there is only one general. A well-run kitchen has a place and a job for everyone. You will rise in position if, and only if you have mastered that position, and can take on more responsibility. A good Chef will know when to give you more responsibility. When you question the Chef, it is considered to be insubordination, and will not be tolerated. People have been dismissed for that. Previously we have talked about knowing your place, and humility, and no matter how great a student you were, you still have much to learn, and being disrespectful won't get you far. You may not realize this, but your colleagues will not hold you in great esteem if you are disrespectful.

1) Does your opinion matter, and in what circumstance would it matter?

_____
_____
_____
_____
_____
_____
_____

2) Is it honest to have an opinion and not express it?

_____
_____
_____
_____
_____
_____

3) Do you see another "kitchen model" that is more democratic? What would it be?

_____
_____
_____
_____
_____
_____

What is Your Definition of Success?

As a culinary student, and would be team member, you have probably been thinking about where you will be working after you get out of school. What kind of food service position will you be looking for? Few if any of you will immediately have a position as chef. If you have learned as much as you can, and practiced everything until it is a reflex, then you have indeed been successful as a student. Understanding the realities of the professional kitchen should help you re=balance what you view as success. We are not here to force feed you our definition of success, but would like you and your class together with your chef/instructor to spend time talking and thinking about how everyday activities can bring you closer to your goal, and how listening, and asking the right questions will move you closer to your goal.

Goals are merely guideposts on your culinary journey. You may feel that you'd like to open your own Italian Bistro that serves food from a little known Italian region. No one is cooking that kind of food, not even the television greats. Your family is willing to help finance this venture, sounds great? But, your only food experience is culinary school, and your work experience is limited. You may think this is a formula for success, but it is really a formula for disaster.

You need to take some time and honestly evaluate what you are good at, and what needs some work; maybe your knife skills are weak. Make yourself the best you can at this stage, and then decide what your goal will be.

1) **What starting position would make you feel successful? Why?**
_____
_____
_____

1) **If you've been working while in school, what are you hoping will happen after you graduate?**
_____
_____
_____

3) **What would you bring to the classroom, practice or your current job now, as opposed to what you brought before?**
_____
_____
_____

4) **How will you feel if others in the industry do not feel you are ready for a lot of responsibility?**
_____
_____
_____

5) **What is the one thing that would make you feel successful?**
_____
_____
_____

Personal Assessment Review

1) What are your broad culinary goals?

_____
_____
_____
_____
_____
_____

2) Have you clearly defined your objective: French restaurant, caterer, or haven't you thought that far ahead?

_____
_____
_____
_____
_____
_____

You need to take some time now to think about why you are in culinary school, and where you see yourself next year, in five years, and in ten years.

3) Why should you be selected for the team?

_____
_____
_____
_____
_____
_____

4) If you have a specific goal, what do you need to do to achieve those goals?

_____
_____
_____
_____
_____

5) What are some of your shortcomings that could stop you from getting to your goal?

_____
_____
_____
_____
_____

6) What skills do you have that help you achieve your goal?

_____

_____

_____

_____

_____

_____

7) What should you do if you have difficulty answering the above questions?

_____

_____

_____

_____

_____

_____

8) What resources are available to you, that could help you answer these questions?

_____

_____

_____

_____

_____

_____

Growth & Development

1) Be true to yourself, evaluate your skills, and be positive.

2) Learn to be as consistent as possible, always.

3) Learn to grow from criticism.

4) Understand that practice will not only make you perfect but also a professional.

5) Understand that in humility you will gain respect in your profession, and also be in tune with everything that is happening in the field.

6) Keep your eyes open, learning can occur in many places.

7) Have faith in your abilities as you continue to read the pages on Competition and Team, and think about how the concepts apply to you.

8) Seek out a mentor.

9) Purchase a 3D Classic Knife cuts Model set.

Competition

Reality of Discipline in the Kitchen

Discipline – in not so many words it is the Internal Control over your desires and impulses and the ability to control and to apply this to reach your goal to be successful. It is also the process of a means to reach a defined conclusion.

In my experiences of the military and the rank structure of the military I can tell you that is very similar to the Culinary Brigade system. There is only one Chef, but there may be other Chefs and or cooks that report to and work for the CHEF. The word Discipline in the kitchen really boils down to everything that you do and how and when you present yourself, appear in a crisp Chef Coat and how well you do your Mise En Place to execute the simply task in the kitchen to create each and every dish. Is it the very best you can do and is it the best tasting and the best looking and the made from the very best wholesome foods that are available the minute the order comes into the kitchen????

No matter how well you work or well you do, sometimes that is not enough. I can tell you this from experience because in Nov 2006, at the Culinary World Cup in Luxembourg, we, The United States Army Culinary Arts Team had all the Discipline in the World and it was well displayed in this arena. But what we did lack was experience as a team and even though it was not displayed there we won a total of 12 gold medals and finished 4th place overall in the world. That is just not good enough!!!! There is a time and a place where you can be too disciplined. It does not matter if you are doing one plate or 1000 plates you must practice perfectly to achieve perfection.

Discipline is one of the daily values that I live by and cook with and conduct myself. It is one of the most important things in a kitchen or in our industry to be successful. It is not an option....You must have a great sense of discipline to achieve your goals, no matter if it is hourly, daily, weekly or monthly goals.

A well Disciplined Chef is always a "Happy Chef" who puts his best food forward and is on time and organized doing so which makes his or her customers very happy which makes the recipients or the guest having a desire to come back over and over again because they are disciplined and know what they want and what they expect.

Chef Rene Marquis, CEC, CCE, PCEC, AAC
SFC, United States Army
Team Captain, USACAT
Military Regional Director, American Academy of Chefs
USCENTCOM, Macdill AFB, Florida

Competition

Competition is a defined activity where competitors seek to achieve a time specific goal, or an objective that is deemed to require above average ability, and be worthy of effort. A spelling bee is a competition, vying for a managerial position is a competition, and the Olympic Games embodies the spirit of competition. Some competition involves individuals striving for excellence, and others pits teams against each other, in order to seek out more and more refined execution and delivery. In the culinary world there are examples of both kinds of competition. From reality TV shows that pit contestants against each other, to team culinary challenges where the team works together, so that the "winning" competitor may be chosen. In ACF competition, the team works together to win or lose.

It is crucial for you to understand that you will be undertaking a team event. You will be part of a well-oiled machine that either wins or loses. Team A wins, which means team B lost. This very simple concept is so important to the teams and your success, that we are belaboring it. It doesn't matter how well you performed if your team lost, because you still lose. In the last section of this book we will discuss and present concepts, concerning teamwork and your job as team member.

Competition will make you stronger, will highlight your weaknesses, and if you are open to the experience, will teach you a lot about yourself, and your future possibilities.

1) Have you thought about and discussed ACF competition with your family and classmates?

_____
_____
_____
_____
_____

2) What do you think is the best thing about competition, and what is the worst?

_____
_____
_____
_____
_____

3) Do you think your career in the food business will be hampered if you don't compete?

_____
_____
_____
_____
_____

4) Do you think competition will bring you a better job?

_____
_____
_____
_____
_____

Never Compromise

Chef Frank Leake once told me the two most important words I will ever take to heart in my career. Those two words were, "never compromise." Since then, whenever the easy road or the less painful option presents itself to me, I recall those two words that prevent me from settling for the mediocre. Every conscious decision I make has behind it a face from my past, which calls to me to make the right decision. I am the product of the sum total of individuals whom I have met throughout my life. Within my food is the mark of every great chef that has shared with me his or her passion to create. True inspiration is the desire within me to create as they have created- to understand that my food is not only an extension of myself, but of every chef that has taught me the joy of creating good food for all who appreciate it. Without inspiration, we would not find the drive within ourselves to achieve the standard our predecessors have set for us. I am aspiring to become a chef/instructor to inspire future chefs to meet and supercede the standards, and to above all, "never compromise."

Lance Nitahara
Candidate, Bachelor's Degree, Bachelor of Professional Studies
Culinary Institute of America
Hyde Park, NY

Should You Compete?

Competition is not for the faint at heart, the easily (emotionally) hurt, those that are closed to learning and new experience, and those who do not like hard work.

ACF competition is about the team. You have to be able to see yourself as part of something, not apart from the team. Teams have names or sobriquets that mean something to everyone on the team. Even non-sports fans know the name Dodgers, and Giants. Sure there is uneven talent on each team, but they work together to get the job done. The first thing you need to ask yourself is whether or not you want to take on more work? How do you know how much work is involved? Ask your coach, ask former team members. Really ask pointed questions, such as: How many hours a week did you devote to preparing for competition? And you should also ask, how did it feel when the team lost? You need to think about how you will feel if you devote next year to preparing for the competition, and your team loses.

1) **Write down why you think you will try out for the team this year.**

_____
_____
_____
_____
_____
_____

2) **Do you think you have to be on the team to further your career?**

_____
_____
_____
_____
_____
_____

3) **Do you think being part of a losing team will adversely affect your career?**

_____
_____
_____
_____
_____
_____

4) **Can you work under pressure with little reward?**

_____
_____
_____
_____
_____
_____

Making the Decision to Compete

Deciding to compete should be a tough decision. If you have suddenly decided to compete without carefully weighing and understanding what you are getting into, you may be unpleasantly surprised.

The first thing you should do is to find out from previous team members and the coach if possible, just what is involved in preparing to compete. If you think you are ready to compete today, you are mistaken. Even if you have terrific knife skills, and you are a plating artist, you need to be able to work in a team environment. We want you to be able to evaluate not only what skills you need to bring to the table, but fully comprehend what life will be like until competition day. You need to find out what the 'organized' practice sessions entail, in terms of time and activity. You will need to learn how to be a team member, and most importantly you have to realize that the time commitment can be very onerous. You will be tired, you will be mad, you will hate your team members and the coach, and you will forget why you wanted to do this in the first place.

1) What do you think you need to learn about yourself before you decide to compete?

_____
_____
_____
_____
_____
_____

2) What skills are you bringing to the competition?

_____
_____
_____
_____
_____
_____

3) How will you get through weeks, and months of preparing to compete, and still have that feeing that you haven't accomplished anything?

_____
_____
_____
_____
_____

4) Have you spoken to previous team members? What have they shared with you about competing?

_____
_____
_____
_____

Consistency in the Professional Kitchen

Consistency is by far the most important hallmark of any business; it defines to your customers who you are and what you stand for. The commitment to consistency is not limited to product or service, but is required in all aspects of ones business, and for that matter, ones personal life. Consider the importance of consistency in decision making, establishing policy, customer service, purchasing, and anything else business related, and the effect of that consistency on those that you do business with.

Consistency is the heart of ones personal and professional integrity. Consistency and integrity go hand in hand in defining ones character. It is setting a standard and never turning ones back on that standard, no matter the cost or the outcome. It is a commitment to your customers and constituents.

Consistency can be defined as the absence of contradictions, and requires a disciplined commitment to core values and beliefs. Strong leadership requires a consistent focus to goals and standards and the discipline to follow through on ones beliefs and values.

The professional kitchen is like a microcosm of any business, it is the receiving department for the business, it is the design facility for the business, it is the production facility for the business, it is the distribution center for the business, and it is the quality control department for the business. It is vital that consistency be maintained throughout all of the processes in the kitchen. Commitment to consistent quality of product and preparation will define your success.

Chef Roy Yamaguchi
Chef/Founder
Roy's Restaurants
Honolulu, Hawaii

Competing Means Learning to Organize & Prioritize

After some soul searching and listening to input from classmates, family and the coach, you have decided that competing is in your future. The next step is to become more organized. Find a system that works for you, whether electronic or paper, and plan how you will keep track of everything in your life. Competing will be a big part of your life, and you cannot just squeeze in the time for it. Preparing to compete will be a big part of every day until your team wins a medal or doesn't. Not everyone leaves competition with a medal. Don't assume you and your team are great, but assume there are other teams out there that are better than your team, and they are your immediate competition for the gold.

Maybe organization is a foreign concept to you, then we suggest you embrace it now, and concentrate on scheduling and keeping to the schedule. If you have practice this afternoon, you can't go the movies instead. Getting everything you need to do in this life isn't easy, adding competition to the mix makes things harder for you, but if you start prioritizing and getting used to a schedule you will succeed. Being exhausted, forgetting to pay the electric bill, and fighting with your friends due to sleep deprivation has happened to former team members. If you don't want to see yourself falling behind in your schoolwork, and hearing that your onions still aren't right after weeks of team practice, you need to make the most of all the organizational tools you can find.

Now is the time to take stock of all that you need to accomplish on a daily, weekly, and monthly basis, and start filling out your calendar. You can create a plan that works for you. If you don't plan for competing you will be too tired to get the most out of practice and preparation, and you will let your team down.

1) How are you organizing your time now? Do you keep everything in your head, and hope for the best? Do you have a planner, but forget to check it everyday?

_____
_____
_____

2) What do you think is the hardest part about starting to get organized?

_____
_____
_____

3) Have you tried to get organized before, but it didn't work, and so you think organization is just not for you?

_____
_____
_____

No matter what you did or didn't do before, once you commit to competition you will have to learn to get organized. Until you find the organizing tool that works for you, life will be exhausting, and you will not be prepared for Success.

Winners Don't Just Happen!

If you've decided that you'd like a spot on the team, you need to prepare yourself mentally, and technique-wise to win. Competing isn't just about winning; it's more about learning about yourself, your team members, and other competing teams. You need to weigh strength against weakness, and learn to decipher positive actions. Winning a medal is a small part of competing, but if you don't think your team has a chance for the gold, you may as well not compete. Being mentally fit to compete means you have the right attitude, you understand your place in the team structure, and you have committed to learning as much as possible to do your part to get the gold. Seeing how you and your team work together to get to the goal, or don't work well at all will teach you many things, that you will use in your career.

You need to be ruthless in identifying what you're bringing to the team, as well as looking at your weaknesses. Correcting these things now will help you later. This is the time to learn about yourself; how you react to adversity, how you react when team members let you down, or when you let the team down. Everything you read in part one of this book comes into play now, as you learn to evaluate yourself, and see how to better yourself. Remind yourself periodically that competing isn't about winning, but about preparing to be the best at all times, and a preparation for the realities of the food industry.

1) What do you feel you need to learn at this stage of your career?

_____
_____
_____
_____
_____
_____

2) You realize that you're the only one on the team who has restaurant experience, yet you're not the team captain. Why do you think that is? What qualities do you think the team captain should have?

_____
_____
_____
_____
_____

3) Why do some teams consistently win? What qualities do the team members have?

_____
_____
_____
_____
_____
_____

Growth & Development

1) Understand your motives for wanting to take part in Culinary competitions.

2) Learn how to marshall all your resources in order to stay focused on your goal.

3) Prepare a personal goal plan for winning. This means understanding the competition requirements, and your place in the team dynamics.

4) Take care of all the little things, like clean uniforms, perfect mise en place.

5) Talk with previous competitors and ask questions. Learn from their experience.

6) Understand that the bar can always be raised, and that you should strive for your personal best.

The Team

What's In It for Me?

If you want to be on the team, you are probably thinking that it is a good move for you. It would be unusual if you were thinking, I want to better the reputation of my school, or coach, or even that you feel your skill set and knowledge can help the team. You're thinking that there is something in it for you, and that's fine. What is important at this point is for you to understand what you will contribute and what you will reap.

You've already read about standards, and what you can contribute to the team, and you've thought about success, and failure. Your class has discussed "Chef Action", and "Being There", and you still feel confident that this is the right step for you. If you truly understand that even winning the gold doesn't mean you've reached the summit of culinary excellence, you may grasp why you would want to be on the team.

What's in it for you is the opportunity to see what you can do. This also means you will see what you cannot do well, and all this will be closely scrutinized by your coach, your team and the judges. Your journey of self-discovery will be made "on-stage" with others watching and judging. We discussed respect earlier, and competition is going to be an arena where you garner or lose respect. You will see that the opinion of others does count, maybe even more than your own.

You should understand that it is good that you want to better yourself, and learn, and improve, because competition is about learning. There will be a bright light shining on you from day one of practice until the actual competition and beyond. If your knife skills are weak, you will not be able to hide that. If you don't work well with others, everyone will also see this. What's in it for you is a chance to grow, and learn, and understand where you stand in terms of competence and artistry and talent. Each competition will make you stronger, and more self assured, because you will have learned how to harness your talents, improve your skills, and concentrate on what is important, the food. In the end, the food is judged. Your wishes and hopes fall to the wayside, because what is on the plate, and what the judges taste is what matters. Competition is a journey of self-discovery, are you brave enough to embark on this trip?

1) **Do you feel that your inspiration and ideas should count more than the execution?**

_____
_____
_____
_____

2) **Do you think being penalized for under-cooked chicken is harsh?**

_____
_____
_____
_____

3) **Do you thinking the judges criteria is open for discussion or outdated?**

_____
_____
_____
_____

Skills

When we say the word chef, the first thing that comes to mind is the uniform, and the second is usually the knife kit. We expect to see clean hands, clean jacket, and no stray hair. We require a " state of cleanliness" in your work area. We expect attention to detail, and we expect you to do everything the way you were taught. This includes not wiping your hands on your apron, not flinging towels over your shoulder, not having sloppy mise en place, and not having small wares piled high in the sink, but rather cleaning as you work . Clean, orderly, on time, and being there ready to use all your senses, is the minimum required.

Knife skills are to be taken seriously and practiced. When you slice an onion, whether in your mother's kitchen or at work, or when you are competing, it should always be the same. Attention to detail, and the standard is what is expected. Plebian onion soup rises to epicurean standards when the classic standard is followed. That means you slice, and caramelize, and cook according to standard. A chef elevates his ingredients by respecting technique and ingredients, anything else is not acceptable.

Your skill level, such as it is, will improve. With time you get better, you can always improve. But you can't take a day off from your knife skills. You can't chop haphazardly at home, and expect to perform at the highest level when competing. Even when there are no judges present there is only one standard. The standard applies when you chop onions alone, when you are in class, or when competing. Flawless execution comes from skills that are so deep within you, that they become you. Great chefs don't have sloppy, messy days, and some good days.

1) **Your instructor says your knife skills are passable, should you be concerned, and what should you do about this?**

_____
_____
_____
_____
_____

2) **Maybe your knife skills are not the best on the team, but everything you cook tastes good. Is this good enough?**

_____
_____
_____
_____
_____

3) **Have you ever read a restaurant review where the critic says the second visit was disappointing, and not as good as the first visit? To what do you attribute the change in quality? What could the chef do make sure this doesn't happen again?**

_____
_____
_____
_____
_____

What Do You Bring to the Team?

Since the team wins or loses as a unit, should you care about your contribution? Being selected for the team means you have a responsibility to your school, your coach, and your teammates. Everyone expects you to be the best you can be, and then to reach further than you think you can. You have to have talent, and drive and perseverance, and most of all be dependable. You have to be there, and give everything to your team. You have to give every day, whether you are tired, haven't had enough sleep, or if you want some R&R with your buddies. Being on the team, means you live and breathe team even beyond the competition. If your team worked like a real team, those people will have a connection to each other forever. Soldiers have a connectedness to each other forever; people who work for corporations where they achieved results as a team also have a connectedness. This connectedness doesn't come from proximity, it comes from mutual respect.

If you aren't committed to bringing your best and more to the team, you will only be going through the motions, and your team will know. Everyone has strengths and weaknesses, if you are honest and humble, you will bring your talents to the team without holding back. If you just go through the motions, and let others work, you cheat yourself and the team. Everyone will know you shirked your duty, and you will lose his or her respect. The culinary world is big, yet eventually everyone knows or knows about everyone worth knowing.

1) What are you bringing to the team?

_____
_____
_____
_____
_____

2) What work habits do you have that will help your team, and what work habits do you have that could compromise success? What can you do to correct this?

_____
_____
_____
_____
_____

3) What should you do if you don't know what you will bring to the team?

_____
_____
_____
_____
_____

4) Should you care what your teammates think about you?

_____
_____
_____
_____
_____

Soigné

Although I worked at a French Cooking School in New York City for five years, first as an instructor and then as the Assistant Director of Education, I do not speak French, nor do I have any formal French training. Therefore you may find it funny that I chose the French word soigné as the word, which best sums up my approach to food and cooking, in particular cooking in a professional kitchen.

My friend and business partner, Drew Breen, first introduced the word soigné to me. An amazing chef and educator in his own right, Drew learned the word while studying at the Culinary Institute of America. He told me his chef would yell it out several times during service. He explained to me that it meant to do it "with care." I thought to myself that the sound of the word itself conjured up visions of sophistication and elegance.

Several years later working along side Drew in our own restaurant in Hawaii, he resurrected the word. I have since passed on the word; but more importantly the concept, to those I have had the good fortune to work along of, and those whom I have taught.

Anyone can make a soufflé, a good soufflé takes time to master; the perfect soufflé is made with love by someone who cares; soigné.

Chef John Richard Akana
Assistant Professor
Hospitality Management Department
City University of New York, College of Technology
New York, NY

Caring

Caring seems like such a touchy-feely concept, especially since we've talked about winning, and discipline, and other hard charging concepts. Caring isn't wimpy; it's what defines you. Winners who don't care seem a little less human to us. We've often heard negative comments made about seemingly successful people because they just didn't seem to care.

What exactly do we mean by caring? We've talked about respect for the ingredients, and your teammates, and of course yourself, but caring means that and a lot more. Caring means you've gone the next step, caring means that you don't send accidents out of the kitchen dressed up in foam and parsley. Probably most customers won't know the difference, but for the one that does, he will feel that you don't care. You may be saying that you can't afford to throw food away, and we're saying that you can't afford not to. Having been the recipients of food that was prepared carelessly, and served by someone who didn't care we were saddened. The waste of food saddened us, we knew that the chef didn't care, and we were also saddened that the chef thought we wouldn't care or notice.

There is no room for mediocrity when you compete; you have to care very much about everything, everyday, all the time. Caring cannot stop, not even for one minute.

Caring means you will take responsibility for things that you have not done.

1) Can you point to a time where you didn't care enough in the kitchen?

_____
_____
_____
_____
_____

2) What should you do, or can you do if those around you don't care about what they are doing?

_____
_____
_____
_____
_____

3) Have your instructors commented on your performance being a little lackluster, a little off the mark?

_____
_____
_____
_____
_____

4) Do you think your work doesn't "shine" because sometimes you only go through the motions and you really don't care?

_____
_____
_____
_____
_____

Constructive Criticism

Being criticized in public, or even in earshot of others isn't easy. We understand that, but in competition everything you do is scrutinized and critiqued. Nothing will escape the judges, and you cannot hide, camouflage or disguise undercooked chicken, sloppy plating, badly thoughout menus, bad technique, and tasteless dishes.

It is understood that you worked in a team environment, but ultimately each team member is responsible for the whole. Your team has lost points due to undercooked or overcooked food. The plating was beautiful, and until the fork hit the food, no one was the wiser. Except your team and you should have known that food was dangerously undercooked, or unappetizing over cooked. Hearing negative comments after you have worked so hard is not easy, but you must learn to listen to it, and take away from it what you need in order to become a winner.

The judges are not cruel and sadistic, and the measure they use to critique your presentation is the same around the world. Your undercooked chicken will be considered undercooked in Munich as well as Paris and London. The standards are the same, and you need to understand that beauty on the plate will not outshine or overcome bad technique.

Balance in taste is also not up for discussion. Classic rules are as valid today as they were 100 years ago.

As soon as possible after competition the team should get together to discuss what the judges said. Write down the judges comments during their review with the team, and later as a team, review each item, noting what went well, and what didn't work. By all means do not gloss over the good. Examine what was good, and understand why, then go on to what didn't work so well. We can all learn from what we did well, but the best teaching tool is when you understand how to make it better.

Losing isn't easy, so lick your wounds, and then figure out what went wrong. The judge's comments are your best guideposts for future success.

1) When you are critiqued, do you listen carefully and ask questions, or are you mad and embarrassed and try to block out what is being said?

_____
_____
_____

2) Judges are impartial, they are telling you the truth, and are holding you up to a standard. What do you think your demeanor should be?

_____
_____
_____

3) Your plans are to open a family style restaurant, and you think your customers are just interested in good taste, and a price point, so why should you care what the judges think?

_____
_____
_____

Stamina in the Kitchen

This year I celebrated my 50-year anniversary of my time as a Chef. Throughout my many experiences in the culinary world, the fact remains that a Chef must have a certain extra something that enables him or her to last in this business. The hours are long, your shifts are spent on your feet, in motion with hot items and fast moving coworkers, tempers can flare, and a mere mistake can throw the whole service if you let it get under your skin. Of course, the profession isn't all bad or people wouldn't be lining the streets in attempt to be a contender on a reality show in a search of the next top chef. However, to be successful and reputable chef you must have the passion to cook, the ability to put out great dishes time and time again and the stamina to survive all the twists and turns that will come day after day. The adage in the culinary world is that you are only as good as your last dish, and this is a powerful and true statement.

While you may think you are a new brand of Super Chef after you complete your first double shift or reach the end of a grueling service, you must confidently stand by each dish that you put out to the public – each dish is just as important as the next. You never know which customer might be the food critic or the socially connected Foodie that will ultimately ruin or glorify your name to his or her friends. One of the best lessons I can pass along from my culinary career is my belief that while stamina is an undeniably important characteristic for the restaurant industry, so is consistency and passion for cooking; a culinarian must develop all of these disciplines philosophies simultaneously to be successful.

Chef Hartmut Handke, CMC, AAC
Chef Proprietor
Handke's Cuisine and Catering
Columbus, Ohio

Energy

When I was asked by Frank Leake to contribute to the "Coaching Culinary Champions", the first word that popped into my mind was ENERGY.

There are many forms of energy cooks / chefs use in the kitchen. They exert physical and mental energy through the rigors of their daily routine, the utilities they use in the form of electricity and propane, but the most outstanding and most often never mentioned is the internal energy force they exude in producing a particular recipe that comes from their hearts, souls and passion. This is the meat of what needs to be taught in all culinary schools. Yes, you require knowledge be it taught or innate to the individual, talent is a nice attribute, but cooking form the heart is what makes the difference in how a dish not only tastes but is perceived by the consumer. Example, have you heard that cooking with a bad attitude spoils the pie? Well, it not only spoils the pie it spoils anything prepared from a cook/ chef who is not in the right energetic space when they enter the kitchen. The mental preparation of the cook / chef is every bit as important as the product used in the recipe being prepared. I will give you this challenge, the next time you come across a meal that is "off" in anyway, take the time and ask the server what mood the cook / chef is in. I will guarantee if the meal is missing anything, it is missing the Energy of cooking from the heart. Positive, passionate, energetic, presence always produces an excellent outcome, whether it is a simple hamburger or 5 star dinner. So I stress to all of my staff that cooking with a smile is the way to become a Culinary Champion.

Chef Amy Ferguson
Proprietor
O's Bistro
Kailua-Kona, Hawaii

Energy

Competition is tough work, and in order to compete well, you need to have high levels of energy. Energy comes in two formats, physical and psychological. You need to be energized physically and mentally every day at practice, and you need to give that last burst of everything you've got at competition.

Lack of planning, and bad skills cannot be overcome with a lot energy; you'll just do bad work at a faster pace. During your practice sessions, both organized and impromptu you need a certain level of energy just to make it through the day. At competition high energy and adrenaline flow will get you through it, but energy cannot make winners. Everything needs to be in place in order for energy to get you over the line, and into the winner's circle.

Physical energy can be enhanced by proper nutrition and adequate sleep. Psychological energy can be enhanced by knowing you are prepared, by visualizing every step of the competition, and by not leaving anything to chance. We've talked about getting organized, from clean uniforms, to organized knife kits. Now you and your team need to learn how to harness this energy to organize your group competition event. Often lack of preparation can lead to lack of energy; because you know your chances are limited you have little enthusiasm and just go through the motions. The best ways for you and your team to create winning energy is to be sure to be prepared as well as possible.

1) After dicing onions week after week, you feel your enthusiasm draining. Why do you think this is happening?

_____
_____
_____

2) Competition is three weeks away and you can't keep your eyelids open during practice sessions. What do you think is wrong? Can you do anything to correct the problem at this point?

_____
_____
_____

3) You forgot to wash your uniforms, and are wearing an apron to hide some stains. Your life is out of control, and any energy you may have had is gone, is there anything you can do to reve up your energy before the competition date?

_____
_____
_____

4) Are some team members still full of energy, while others are winding down? Are some team members grumbling about practicing when they could be catching up on sleep? How are you feeling right now, full of energy and excitement, or tired, listless, and ready to throw in the towel?

_____
_____
_____

Friendship

Friendship-Team-Ambition becomes interconnected as a young cook begins learning in the industry. The balance of strengths and weaknesses among your friends enables a trust among individuals, growing into a team of people who begin counting on each other, growing together in their respective ambitions.

Although most young cooks and chefs do not develop as "friends", there has always been that line not to be crossed. Friendship is substituted with respect. And although friendship may not always develop with the chef, usually there is a series of friendships with those directly connected to the chef; the sous chef, chef saucier, etc. This was the case in my own apprenticeship, the saucier and I became friends, I became a part of his team, mentoring helped establish and develop my ambitions, my career was born. All this began with friendship.

Friendships are found and learned best through travel and life experiences both on and away from the job. As you train together, you grow closer, there are common goals. Learning both in school and through hands on experiences helps in the development of technique and in sustainability of the classics. Learning the classics are essential to whom you must become as a cook and chef.

No matter your position in a restaurant, it is not a one man or one-woman show. As the leader of a team it is important that the team never misinterpret that role or their individual roles and never misunderstand a possible friendship connection. Most chefs forgo the friendships with their employees in order to maintain that understanding. A side note, Frank was only one of a few friendships that developed when he was part of my team. A trust and understanding of each other's roles developed and the friendship remains years following his departure from the team.

Finding friends among your peers at work could be the key to career development. The friendship interconnects with those of your team members; you find common grounds, interests, standards and ambitions. Suddenly you find your chosen profession as a chef taking off. You are on your way and you are in good company, sharing those ups and downs with that team that you work with.

Chef Rolf Walter, AAC
Executive Chef
Hale Koa Hotel
Honolulu, Hawaii

Team Dynamics

You are now a team member. Congratulations, enjoy the fleeting feeling of triumph and glory, and get set to work like you have never worked before. Once the coach puts the team together, it is expected that you all will work together for a common goal. You will all win or lose together. There is no personal best, no personal glory. If the chicken is raw, and even if you didn't cook it, you still lose.

Respect

You have to respect yourself, and all the team members. Don't forget the coach because any lack of respect on your part toward him/her diminishes the whole team. You have to assume that every team member has the same goals, and abilities. If you feel superior you are damaging the team. Everyone has strengths and weaknesses, but together you can win. This is not a one-man show, and you need to work hard within the team environment.

Criticism

Expect to receive a lot of constructive criticism, and receive it and use it to your advantage. The better you become the better the team is. The coach will be monitoring everyone, and you and your teammates should be watching each other. A team member, who knows better, should gently alert his team members to concepts or actions that could be better accomplished. All team members need to watch out for each other, and make sure that everyone is as good technically, and artistically as possible.

Mentoring

The coach will be mentoring and advising as the year progresses, but every team member who has a strong skill or talent should be sharing their abilities with the other members. Sharing with teammates will not diminish you; you will only have a better chance at winning at a team event. Everyone has a different learning style, but if you see that your teammates tourne potatoes are not perfect, you need to invest time to help them learn how to reach the level of excellence required. Teamwork means everyone works to their abilities for a common goal, and those who can do better have to assist those who need help.

Recognition

We all like praise, but there will be little praise directed at individuals in this scenario. Whether or not your vegetables are perfect or not the team gets the praise, and the criticism. The team will be recognized for achievement, or failure. You need to be sure of yourself, and understand that this is not about you, but about the team. When you notice team members accomplishing things, you need to acknowledge that. The team that respects, supports, and works well together is the team that wins. No one said you have to like your team members.

1.) What can you do to make the team stronger?

_____

_____

_____

_____

_____

_____

2) What weaknesses do you have that could bring the team down?

_____

_____

_____

_____

_____

3) You feel that other team members don't respect your opinion, what should you do?

_____

_____

_____

_____

_____

Pivotal Key for Success

Short of quality parenting for most, a highly skilled, well balanced mentor, coach or teacher who is full of wisdom, regardless of the subject/skill, is the pivotal key for success in any industry. Understanding that concept, we should also all understand that time is our most prized possession and in the moment it is finite. Use it wisely!!!

Chef William C. Franklin, CMC
Corporate Executive Chef
National Account Manager
Onsite FSM
Nestle Foodservice North America
Centennial, Colorado

Time

Time is an important concept to consider at this point. Getting to practice on time everyday, and accomplishing all tasks in a timely manner all the time no matter what happens, is essential for winning.

No matter how well your product turns out, if it isn't done on time, you lose.

Practice sessions should always have a time component, and someone needs to be a timekeeper. The clock is always the enemy, but you can make it your friend if you get organized, and schedule tasks. You won't get in the weeds if you can choreograph your "performance" to a timed schedule. If the team learns how to work with the clock instead of against it, your competitive events will run more smoothly, and you will be able to concentrate on your performance.

1) How can you learn to work with the clock instead of against it?

_____
_____
_____
_____
_____
_____

2) As a team member what is your responsibility toward keeping other team members on target and on time?

_____
_____
_____
_____
_____
_____

3) You feel that cooking is a creative expression, and you don't like to work within time constraints. How can you still be creative and get your work done in a timely fashion?

_____
_____
_____
_____
_____

Cleanliness is Excellence

Cleanliness in a professional kitchen is paramount and goes beyond what is required by local or federal health codes when considering what it means to promote excellence. It is a belief, a core value. Cleanliness, like other mise en place that is done in a kitchen comes down to the creation of a culture that promotes total professionalism amongst those working within the environment; cleanliness with regards to equipment, personal hygiene, and so on. For example, should a chef's coat get soiled, the chef would change it, or at a minimum revere the flap, so as to present a clean coat front. This is vital especially in situations where chefs are working more and more in exhibition-style kitchens. Chefs are on stage, and customers can observe both the good and bad practices. Cleanliness is not only a practice, but also an attitude that carries with it an image of professionalism and promotes the overall respect for the profession.

Chef Jill Bosich, CEC, CCE, AAC, CFSP
Executive Chef, Energy Programs Advisor, Southern California Gas Company
Downey, California

To Be a Team Member

To be a team member in a professional kitchen, an individual must be a life long learner with a positive attitude. This is someone who can maximize his/her contribution and push the team forward to fulfill its mission.

Chef Jon T. Matsubara
Chef/Entrepreneur
Honolulu, Hawaii

Pride in Ownership

During medieval times artisans signed their work. They worked hard, were eventually admitted into guilds, and always signed their work. They were proud of their accomplishments.

Are you always proud of your product? As a member of the culinary team you must be proud of what the team produces. You can't say my work is good, but my team member's work falls short of excellent. The team is judged as a whole, would you sign your name to what the team presents to the judges? If you are horrified to think that your name would be linked with your team members, and after competition fervently hope the judges won't remember that you were on the team, you and your team have a big problem to resolve.

You must be proud of everyone's work. You will be judged by what the others contribute to the final product. No judge will say your plating was good, but Joe's chicken was raw, and Mary's vegetables were overdone.

If you respect the food, yourself and your team members, you will never have sloppy diced onions, and under caramelized vegetables. You and your teammates will strive for clean, organized mise en place, efficient and precise movements to complete all tasks, and always feel able to "sign" your name to the work accomplished.

1) I feel my work is of superior quality but some team members don't have good kitchen habits, and lack skills. How can I be proud of our accomplishments?

_____
_____
_____
_____
_____
_____
_____

2) You work hard; keep track of time, yet others in the group keep dragging you down. What can you do to assess the situation and help to makes changes that benefit everyone?

_____
_____
_____
_____
_____
_____

3) You feel that the coach doesn't seem to notice mistakes that are bringing the team down, can you do anything about this?

_____
_____
_____
_____
_____
_____

A Vision & A Mission

Corporations and teams within them each have a mission statement, and a shared vision. Your team must have an articulated common and unified objective. Just wanting to win is not sufficient. Each member must contribute to this vision and mission statement. It must be written down, and all members must have this credo on a card in their pocket at all times. All practice sessions should commence with the team leader clearly stating this mission statement.

Your team should have a name. Not just the red team, or the green team. It should have a name that resounds with meaning for every member of this team. It's like your secret code, and when you hear your team's name it should differentiate you and your team members from all the other teams.

At competition when your team's name is announced it should fill all of you with pride.

1) You are only going to be on the team once, why should you go to the trouble of working on a vision and mission statement?

_____
_____
_____
_____
_____
_____

2) You don't like to feel pressured into compliance, and you're concerned that you will be bullied into accepting a vision you don't like. What can you do?

_____
_____
_____
_____
_____

3) What will happen if your team doesn't have a mission statement? After all you all want to win.

_____
_____
_____
_____
_____

Chef as a Humanitarian

Never underestimate the power of food. As a chef, you should not forget that it is not just about the restaurant. But as you are taught and learn to cook and eventually become a chef your whole world opens up and you are provided opportunities to see and change the world. Your degree will take you as far as you want to go. A chef once shared with me that " it's not about how talented you are as much as it is how much energy you have."

Food is an important part of daily living, for everyone. We must give respect to the very young and the old. We learn in one form or another to appreciate a good meal, and learn quickly as we grow that food always tastes best when shared.

In my own life, during the past five years, food has been a way to bring children together. Four years ago my partner and I founded Common Threads, not for profit, www.ourcommonthreads.org , to teach children about the world through the act of cooking. I continue experiencing the positive affects of cooking with and for children. We have grown from twenty to nearly a thousand children loving to cook in twelve locations throughout Chicago, Mississippi and Los Angele. Cooking has become a very big part of their lives. They and their families get to experience these positive affects at the table.

Everyone loves food, whether experiencing the benefits of cooking at home or in the classroom, food is a great way to teach, learn and heal. Understanding food and cooking helps one to better understand and pay respect to people of the world and their cultures.

Food: the connection—to our past, to our future. When coming together at the table, historically food acts as the means to bring us all together. What is important to remember is whether it's a family event, a holiday party or a business meeting, whatever the occasion, nothing is more appropriate than a meal and sitting at a table and conversing. Nothing brings people together and holds them together by a common thread more than food.

Food is a significant part of the experience, but the act of sitting and interacting and conversing is important when analyzing the power of food. Never underestimate the power of food and the strength it has in all human existence. Throughout history food has and always will be connected to the search for answers and outcomes.

If you have issues with people cook for them, it has the uncanny ability to provide a number of great things. In both good and bad times, food and cooking are always appropriate.

Chef Art Smith
Special Events Chef to Oprah Winfrey
National Best Selling Author
Restaurant Proprietor TABLE 52
James Beard Award Winner
Chicago, Illinois

At the End of the Day

At the end of every practice session the team must take the time to see what went well, and what didn't. You will learn more from your mistakes than from your successes. Don't stint on the time spent for this very important exercise.

This is not a time to vent, or pick on team members, but rather a time to see what processes worked and what didn't. If the soup was held up because someone chopped onions too slowly, the corrective measure is to mentor that person in skills until they can meet the standard. Remember the standard applies to everything you and the team does. The clock is ticking, someone has been assigned to watch the clock, and everyone needs to get their skills and productivity to the same level. That means raising the bar for some, not lowering the bar to accommodate others.

If you remember that competition is about refining a process, and making everything better always, you will see that an end of the day review is crucial for your success as well as the team's success.

You may not feel that what happens during team competition will affect your career, but unless you plan to be a one-man shop, you will always be working with many people. Even if your restaurant has received gold stars, you do not work alone. Learning how to bring out the best in everyone, including yourself is a skill that will serve you well. Learning how to elicit the best from others makes for a strong leader. A good chef is head of state in his kitchen, and must be a good leader, and must be prepared to compete.

Growth & Development

1) Understand the target in competition and teamwork is not winning but growth and knowledge.

2) Learn how to work as a team.

3) Watch each others back.

4) Work and play well together, because you need each other to learn and grow.

5) Remember to think and articulate safety and sanitation every day, all the time.

6) Remember it's always about the food.

7) Identify, and focus on the target.

8) Learn to work with all your senses.

9) Learn how to plan and implement time management practices.

Food is the Star

In this business food will always be the star. You can't forget that. People come to your restaurant to eat. From the short order cook to the Food Network chef, to a Michelin star chef, it all comes down to fulfilling this need. You, the chef, are the vehicle to deliver food to people.

To make food the star you first need to be passionate and have a real love of cooking. The next step is to study, eat, practice and work for several chefs you respect and want to learn from. As an experienced cook you can then use your innate talent, your learned skills, your experience, and your ambition to help you decide what to do with the raw ingredients-how to prepare the food that will represent who you are as a cook, and as a person. Once you have found your own culinary point of view, then you have something to share with your customers.

Not everyone will be a household name, but if you start out with the passion for what you do, a respect for the food, and the belief that the food is the most important thing, everything else will fall into place. If you have a great product, and you have integrity, people will notice.

Chef Tony Liu
August Restaurant, New York City
Competitor/Iron Chef America

In Conclusion

We feel confident that you and your classmates, together with your chef/instructor have learned about yourself, and each other. We know that discussion and communication is important for success. You've chosen a profession that continually seeks excellence, and also remembers that the customer is the one who will decide your future.

A love of food, a sense of humility, and an appreciation for the many possibilities of each ingredient will not only enrich your customers palate, but will also enrich your life.

Wear your uniform with pride, and seek to better yourself everyday. Your destination is also where you start. Good luck!

We are interested in your career path, and welcome your comments on our blog: http:// coachingculinarychampions.blogspot.com.

Aspiration

"If you choose a career in the culinary or hospitality business, always remember that it is a celebration. If you are not hospitable, and if you do not focus on this aspect, the career may not be for you. Culinary and Hospitality offer many wonderful careers, but you must aspire to 100% dedication. Aspire to food as a celebration!!!"

Chef Joe Amendola, CEPC, CCE, AAC
Senior Vice President and Principal of FESSEL International
Ambassador to the Culinary Institute of America

READ:
Chef Andre Soltner's comments on Professionalism.
Chef Art Smith's comments on Humanitarianism.
Chef Roy Yamaguchi's comments on Consistency.

THE AUTHORS
Mareva Lynde is an Information Management Consultant, writer, publisher, and food enthusiast.

Chef Frank Leake, CCC, CCE, '74, Alumni CIA, Professor of Fundamentals of Cookery at Kapi'olani Community College at the University of Hawai'i, and is the former coach of two ACF Culinary teams in Hawai'i.

ISBN: 978-0-9799608-0-2

$23

CPSIA information can be obtained
at www.ICGtesting.com
Printed in the USA
BVOW03s2052110517

483405BV00009B/7/P